BEI GRIN MACHT SICH IHR WISSEN BEZAHLT

AF141592

- Wir veröffentlichen Ihre Hausarbeit,
  Bachelor- und Masterarbeit

- Ihr eigenes eBook und Buch -
  weltweit in allen wichtigen Shops

- Verdienen Sie an jedem Verkauf

Jetzt bei www.GRIN.com hochladen und kostenlos publizieren

Luisa Liebold

Discovering The Calendar

A Project for Bilingual Maths Lessons in Elementary School

GRIN Verlag

Bibliografische Information der Deutschen Nationalbibliothek:

Die Deutsche Bibliothek verzeichnet diese Publikation in der Deutschen National-bibliografie; detaillierte bibliografische Daten sind im Internet über http://dnb.d-nb.de/ abrufbar.

Impressum:

Copyright © 2011 GRIN Verlag GmbH
Druck und Bindung: Books on Demand GmbH, Norderstedt Germany
ISBN: 978-3-656-02941-0

Dieses Buch bei GRIN:

http://www.grin.com/de/e-book/180129/discovering-the-calendar

GRIN - Your knowledge has value

Der GRIN Verlag publiziert seit 1998 wissenschaftliche Arbeiten von Studenten, Hochschullehrern und anderen Akademikern als eBook und gedrucktes Buch. Die Verlagswebsite www.grin.com ist die ideale Plattform zur Veröffentlichung von Hausarbeiten, Abschlussarbeiten, wissenschaftlichen Aufsätzen, Dissertationen und Fachbüchern.

Besuchen Sie uns im Internet:

http://www.grin.com/

http://www.facebook.com/grincom

http://www.twitter.com/grin_com

Pädagogische Hochschule Freiburg
Fakultät für Kultur- und Sozialwissenschaften - Europalehramt
Seminar: Durchführung zielsprachiger Projekte für die Primarstufe
in Verbindung mit dem Sachfach

Discovering the Calendar

Allgemeine Angaben

Name:	Luisa Liebold
Studienfächer:	Englisch, Mathematik, Deutsch
Studiengang:	Europalehramt an Grund-, Haupt- und Werkrealschulen, Schwerpunkt Grundschule

Angaben zum Projekt

Titel des Projekts:	Discovering the Calendar
Sprache:	Englisch
Sachfach:	Mathematik
Schulart:	Grundschule
Klassenstufe:	4

Inhaltsverzeichnis

Einleitung

Obwohl der bilinguale Unterricht in den letzten zehn Jahren immer verstärkter Einzug in die deutschen Klassenzimmer gefunden hat, wird dieses Konzept im Mathematikunterricht noch immer selten eingesetzt. Dabei bietet gerade der Mathematikunterricht viel Anlass zur Handlungsorientierung und zahlreiche Möglichkeiten zur interkulturellen Perspektive. Im Sommersemester 2010 ging ich zusammen mit meiner Kommilitonin ▮▮▮▮▮▮▮▮▮▮▮▮ auf die Suche nach einem geeigneten Thema für ein Projekt im bilingualen Mathematikunterricht der Grundschule. Vor allem der Bereich der Größen und der Geometrie, so zeigte sich bei einer Recherche in der Literatur, bietet dabei zahlreiche Umsetzungsmöglichkeiten. Nach einigen Überlegungen entschlossen wir uns, den Kalender zum Thema unseres Projektes zu machen. Wir wählten dieses Thema, da es durchaus große Relevanz im Leben einer jeden Schülerin und eines jeden Schülers hat, vielleicht aber gerade auf Grund seines Lebensweltbezugs eher selten im schulischen Unterricht beleuchtet wird. Wir gingen davon aus, dass den Schülerinnen und Schülern nicht alle Begriffe bekannt waren und wollten sie im Rahmen unseres Projektes vertrauter mit diesem Thema machen.

Auf Grund meines zweisemestrigen und ▮▮▮▮▮▮s einsemestrigen Auslandsaufenthaltes war es uns nicht möglich, die Durchführung des Projektes gemeinsam zu tätigen. So hielt ▮▮▮▮▮▮ ihre Stunde bereits im März 2011 in einer dritten und vierten Klasse einer baden-württembergischen Grundschule, während ich mein Projekt erst im September 2011 in der vierten Klasse einer sächsischen Grundschule durchführte. Was uns zuerst als Nachteil erschien, entpuppte sich bei der Vorbereitung meines Projektes als Vorteil, denn ich konnte die Erfahrung, die ▮▮▮▮▮▮ bei ihrem Projekt sammeln konnte, in mein Projekt einfließen lassen und so die Qualität des Projektes weiter verbessern.

In dieser Arbeit gehe ich zuerst auf die Bedeutung des Kalenders im Allgemeinen und mit Bezug auf die Schülerinnen und Schüler im Besonderen ein, bevor ich didaktische Aspekte zur Durchführung des Projekts im Unterricht näher beleuchte sowie Lernziele darstelle. Des Weiteren werden die theoretischen Grundlagen des bilingualen Unterrichts, insbesondere des bilingualen Mathematikunterrichts, vorgestellt. Abschließend liegt der Fokus auf meinem Auslandsemester. Im letzten Teil dieser Arbeit findet sich dazu das Portfolio zum Auslandsemester.

1. Sachanalyse

For centuries, calendars have been used to research regularity and to document the lapse of time: "**Kalender** bilden den Zeitablauf ab, um Vergangenheit, Gegenwart und Zukunft zu beschreiben. Ein **Datum** benennt dabei bestimmte, einzelne Zeitpunkte." (Görke 2011, p.3, emph. in original). The way in which days would be counted was defined but different peoples in the world used different methods. Thus, for Slaves and other agricultural peoples, the period of time between two harvests was a year, whereas for American Indians, a new year started after the first snowfall and for Australians after the first rainfall (Sawelski 1977, p. 12). These ways of measuring were very inaccurate (ibid.).

4000 years ago in Ancient Babylon, the first moon calendar measured a year beginning with the new moon (Sawelski 1977). The months were 29 and 30 days long so that there were twelve months with a total of 354 days (ibid.). There was a gap to the real length of a year which is 365.2422 days long (ibid.). Because of this, every three to eight years, an extra month was added.

In 4 BC, the Hebrews introduce a so called Lunisolar Calendar based on the moon and the sun (Sawelski 1977). It has twelve months, each with 29 or 30 days (ibid.). Seven times in 19 years, a 30-day month is added (ibid.). Even after adding this extra month, there was still a mismatch between the length of the calendar and the real length of the year; so another day is added every year that starts with a Sunday, Wednesday or Friday (ibid.). Despite the high number of corrections necessary to adjust this calendar to the real length of the year, in the end it matches the revolution of the earth around the sun quite well (ibid.).

In 5 BC in Ancient Egypt, the calendar used twelve months and each one was 30 days long (Sawelski 1977). Five days were added at the end of each year (ibid.). Although it was known that the year was "moving", nothing was changed because of cultic reasons (ibid.).

During the seventh century BC, the Mohammedan calendar was introduced (Sawelski 1977, p. 15). It was based on the change of the phases of the moon and had twelve moon months; six months were 29 days long; six months were 30 days long (ibid.). The calendar started on Friday, July 16, 622, the assumed date of the Hegira, Mohammed's escape from Mecca to Medina (Görke 2011, p. 46).

In the Roman Empire, debts had to be paid on the first day of every month called "calendae" (Sawelski 1977, p. 16). The Latin word for the debt register was called "calendarium", thus the term used until today (ibid.).

In 46 BC, Julius Caesar ordered a reform of the calendar, introducing the year consisting of 365 days with a day added every four years (Sawelski 1977, p. 16). The calendar is known as the Julian calendar (ibid.). Because this calendar was still not quite exact, Pope Gregor reduced the year 1582 by ten days (Kidsnet.de). For a year to be a leap year, it must be possible to divide its number by 4 (ibid.). This Gregorian calendar is used in most parts of the world today (Sawelski 1977). Today February 29th is added in the leap year while February 24th used to be doubled during the Roman Empire and the Middle Ages (Görke 2011, p.4). In Russia where the moon calendar was used, the Gregorian calendar was not introduced until 1918 (Sawelski 1977). The ISO year is a modern addition to the Gregorian calendar (Görke 2011, p. 22). The weeks of the year are numbered as well. The rule is that "die erste Woche des Jahres enthält dabei stets dessen ersten Donnerstag." (ibid.).

The common underlying similarity of all calendars is that the date is determined by considering the rotation of earth and moon compared to the sun (Görke 2011, p. 127).

Görke (2011) describes the function of a modern calendar: "zukünftige oder vergangene Kalenderdaten einschließlich der beweglichen Kirchenfeste korrekt angeben und andererseits alte Daten korrekt einordnen." (p. 85). Historians have struggled to specify historic dates. For this reason, the so called "Julian day" was invented (ibid., p. 20). Every day is counted continuously and has a different natural number. Counting starts on the first day of 4713 BC (ibid.). From this it follows that Saturday, April 25, 2009 is Julian day 2 454 947 and Friday, January 1, 2010 is Julian day 2 455 198 (ibid.).

During the course of time, different peoples, scholars, philosophers, poets and rulers have influenced the development of the calendar (Sawelski 1977). Scientific, political and religious factors are interwoven in it (ibid.) but the basis for every calendar is the day (Görke 2011, p. 3). Bigger units are week, month and the year; smaller units are hour, minute and second (ibid.). A day has 24 hours that are 60 minutes each; every minute is 60 seconds long (Görke 2011, p. 4). The length of a second is defined by the oscillation period of the cesium atom (ibid.). There is a slight mismatch between the oscillation period of the cesium atom and the length of a so called civic second (ibid.). When the sun passes the meridian at noon, one day has passed. The length of a day is divided by 24 times 60 times 60 which is equivalent to 86400 seconds (ibid.).

Common abbreviations are:

- a – year
- mon / m – month

- w – week
- d – day
- h – hour
- mon / m – minute
- sec / s – second (Görke 2011, p. 4)

2. Didaktische Analyse

2.1. Themenbegründung

Wie schon in der Einleitung erwähnt, haben wir das Thema Kalender zum Mittelpunkt unseres Projektes gemacht, weil es einen starken Lebensweltbezug für die Kinder hat. Darüber hinaus lässt es sich aber auch gut mit den Zielen im Bildungsplan vereinbaren.

Mathematik

Am Ende der zweiten Klasse können die Schülerinnen und Schüler „Zahlen lesen, sprechen und darstellen, [...] Zeit vergleichen, schätzen und messen" (Ministerium für Kultus, Jugend und Sport Baden-Württemberg 2004, S. 58f.). Des Weiteren sollen die Schülerinnen und Schüler in der Lage sein, „Zeitpunkte und Zeitspannen in einfachen Fällen an [...] Kalender bestimmen" (ebd., S. 58) sowie „in einfachen Sachsituationen Zeitpunkte und Zeitspannen berechnen" (ebd.) zu können. Dabei entdecken sie „Zahlbeziehungen und Regelhaftes" (ebd., S. 54). Denken und Lernen finden im „handelnden Umgang mit Materialien" (ebd.) statt. Am Ende von Klasse 4 können die Schülerinnen und Schüler „ihr Wissen und Können im Umgang mit Größen zur Klärung realistischer, kindgemäßer Sachverhalte nutzen; [...] aus Texten Daten sammeln, erheben und darstellen; Sachsituationen und Sachverhalte, die in Bildern, Tabellen und Diagrammen dargestellt sind, interpretieren und mathematisieren" (ebd., S. 60f.)

Dabei wird auch im Mathematikunterricht ein interkultureller Bezug angestrebt: „Die Schülerinnen und Schüler weiten ihren Blick über die Nachbarschaft, die Stadt, die Republik hinaus zu Nachbarländern, zu Europa, zur Welt – sie gewinnen mit der weltbürgerlichen Freiheit einen Sinn für die Besonderheit ihres eigenen Volkes, ihrer eigenen Sprache, ihres eigenen Landes." (Ministerium für Kultus, Jugend und Sport Baden-Württemberg 2004, S. 12). Dass dieses Lernen auch in der Fremdsprache geschehen kann, ja nicht nur anderen Fächern vorbehalten bleiben sollte, wird ebenfalls betont (ebd., S. 56).

Da „Mathematik [...] wechselseitig mit anderen Fächern und Fächerverbünden vernetzt" ist (Ministerium für Kultus, Jugend und Sport Baden-Württemberg 2004, S. 56), kann außerdem fächerübergreifendes und fächerverbindendes Lernen stattfinden.

Englisch

Als vorrangiges Ziel wird die Entwicklung von kommunikativer Kompetenz gesehen

(Ministerium für Kultus, Jugend und Sport Baden-Württemberg 2004, S. 64ff.), wobei „die Einbettung der Zielsprache in Sachfächer als Beitrag zum Lehren und Lernen [...], wann immer möglich, anzustreben" ist (ebd., S. 68). Dabei kann auch dem Anspruch eines situationsbezogenen und authentischen Einsatzes der Fremdsprache zu echter Mitteilung (ebd., S. 72) nachgekommen werden. Die Sprechanlässe sollen aus der Erlebniswelt der Kinder kommen (ebd.). Die Fremdsprachenkenntnisse sollen spiralförmig durch das Einbinden von Bekanntem und das Anknüpfen an Neues aufgebaut werden (ebd.).

Die Schülerinnen und Schüler verstehen im mündlichen Bereich kurze beschreibende und erzählende Texte, kurze Dialoge und Äußerungen und in Klasse 3 und 4 auch kürzere beschreibende und erzählende schriftliche Texte (Ministerium für Kultus, Jugend und Sport Baden-Württemberg 2004, S. 68).

Am Ende der zweiten Klasse können die Schülerinnen und Schüler „ein kurzes Lied nachsprechen, auswendig lernen, in der Gruppe vortragen", sie „können sich vorstellen" und sie „kennen exemplarisch einige Alltagsgewohnheiten und Konventionen aus zielsprachlichen Kulturen", „sind für die Verschiedenartigkeit von Sprache sensibilisiert" und „für die Unterschiede und Gemeinsamkeiten hinsichtlich möglicher Organisationsformen des alltäglichen Lebens sensibilisiert" (Ministerium für Kultus, Jugend und Sport Baden-Württemberg 2004, S. 74ff.). Darüber hinaus können sie am Ende der vierten Klasse „einen kurzen geschriebenen Text vorlesen", „mit Hilfsmitteln eine sehr kurze und einfache Präsentation gestalten" und „einfache Anfragen zu Wissenszusammenhängen aus bekannten Themenfeldern auch auf Englisch beantworten" (ebd., S. 77ff.)

2.2. Relevanz des Themas für die Schülerinnen und Schüler

Spätestens mit Schulbeginn, oft aber auch schon davor, ist es heute üblich, die Freizeit von Kindern zu strukturieren. So planen Eltern außerschulische Aktivitäten für ihre Kinder und sorgen dafür, dass diese regelmäßig Reitstunden, Ballett, Englischunterricht, Klavierstunden usw. besuchen. Auch der Stundenplan ist eine Strukturierung des Alltags der Kinder und stellt einen Kalender im Kleinen mit Fokus auf den Verlauf einer Woche dar.

Zeitmessung und Zeitplanung werden durch die vielseitige Gestaltung des Alltags für Kinder immer relevanter. Darüber hinaus begegnen die Schülerinnen und Schüler Datumsangaben unweigerlich bei Geburtstagsfeiern, in den Nachrichten, beim Verabreden mit Freunden, bei der Ferienplanung, beim Feiern von Festen und vielem mehr.

Das Lesen des Kalenders und die richtige Interpretation von Datumsangaben spielt dabei eine zentrale Rolle. Nur durch das Verstehen des Aufbaus eines Kalenders werden die Schülerinnen und Schüler in der Lage sein, Gegenwart, aber auch Vergangenheit und Zukunft vorstellbar und sinnhaftig zu machen.

Das Hervorheben von Gemeinsamkeiten, aber auch das Offenlegen von Unterschieden im Vergleich zwischen den Kalendern und der Datumsangabe in verschiedenen Kulturen führt zu verbesserter interkultureller Kompetenz und zum Verständnis für Andere. Dabei ist die Auseinandersetzung mit dem Thema auch in der Fremdsprache wichtig und zweckmäßig.

2.3. Didaktische Reduktion

Das Thema „Kalender" ist sehr vielseitig und ist in verschiedenen Disziplinen wie Geschichte, Astrologie, Mathematik, Religion, aber auch Politik von Bedeutung (vgl. Kapitel 1). Auf Grund zeitlicher Begrenzung, aber auch durch die fachliche Fokussierung auf das Thema, beschränken wir uns in unserem Projekt „Discovering the Calendar" auf die Struktur des Kalenders sowie das Lesen und Interpretieren von Datumsangaben und Zeitspannen. Durch sprachliche Einschränkungen als Folge der Verwendung der Fremdsprache legen wir den Schwerpunkt weiterhin auf das richtige Lesen von Datumsangaben unter dem Aspekt der unterschiedlichen Darstellung in der anglo-amerikanischen Kultur.

Das Thema „Kalender" als Teil des Themas „Zeit" wird oft im Zusammenhang mit der „Uhr" behandelt. Aus zeitlichen Gründen und auf Grund der Komplexität beider Darstellungsformen von Zeit haben wir uns auf den Aspekt des „Kalenders" und damit auf die Darstellung der Zeit mit der kleinsten Einheit des Tages entschieden.

3. Lernziele

Durch die Durchführung des Projektes im fremdsprachlichen Sachunterricht werden neben fachlichen Zielen auch fremdsprachliche Lernziele verfolgt. So sollen die Schülerinnen und Schüler zunächst die Struktur des Kalenders wiederholen und festigen und die Einteilung des Jahres in Monate, Wochen und Tage erarbeiten. Sprachlich erweitern sie ihren Wortschatz um die englischsprachigen Begriffe. Das Nennen und Einordnen des Geburtstags in den Jahresverlauf führt die Schüler zum Kennenlernen von Ordinalzahlen in der Fremdsprache und zu einem Bewusstsein für die Unterschiede zwischen Kardinal- und Ordinalzahlen. Auf der Grundlage des Wissens um die Datumsangabe in der deutschen Sprache werden den Schülerinnen und Schüler die Unterschiede in der Darstellung in der englischen Sprache bewusst gemacht. Die Schülerinnen und Schüler sind des Weiteren in der Lage, sich im Monatskalender zurecht zu finden und können ausgehend von einem vorgegebenen Datum die Begriffe „yesterday", „tomorrow", „next week" sowie „... week ago" interpretieren und den Wochentag bestimmen sowie das Datum angeben. Außerdem können die Schülerinnen und Schüler Datumsangaben nutzen, um Zeitspannen – auch über Monats- und Jahresgrenzen hinweg – bestimmen zu können und miteinander zu vergleichen.

Im Rahmen des Projektes werden auch personale und soziale Lernziele verfolgt. So lernen die Schülerinnen und Schüler, allein und in der Gruppe Verantwortung für ihre Arbeit zu übernehmen und an einem gemeinsamen Ziel zu arbeiten. Leistungsstärkere helfen leistungsschwächeren Schülerinnen und Schülern. Das Präsentieren von Lernergebnissen im Klassenverbund ist ein weiteres Ziel dieses Projekts.

4. Theoretische Grundlagen des bilingualen Unterrichts

Bilingualer Unterricht ist in deutschen Klassenzimmern auf dem Vormarsch. Gab es 1999 noch 366 Schulen mit bilingualem Angebot, so waren es 2005 schon 847 Schulen (Sekretariat der Ständigen Konferenz der Kultusminister der Länder in der Bundesrepublik Deutschland 2006). Dabei gibt es in der Literatur noch immer Verwirrung um die Terminologie „bilingualer Unterricht" (Badertscher / Bieri 2009, S. 11). Während die Begriffe „Immersion" und „bilingualer Unterricht" den Akzent eher auf die sprachliche Dimension legen, ist bei „bilingualer Sachfachunterricht" sowie „Content and Language Integrated Learning" auch der Inhaltsaspekt in der Bezeichnung enthalten (ebd., S. 11). Mittlerweile hat man sich auf europäischer Ebene auf den Begriff „Content and Language Integrated Learning" (CLIL) geeinigt: „the acronym CLIL is used as a generic term to describe all types of provision in which a second language (a foreign, regional or minority language and / or another official state language) is used to teach certain subjects in the curriculum other than language lessons themselves." (Eurydice 2006, S. 8). Auch wenn weiterhin Begrifflichkeiten gemischt verwendet werden, so sollte klar sein, welches Ziel mit dieser Art des Unterrichts verfolgt wird: "It [CLIL] is meant to ensure first that pupils acquire knowledge of curricular subjects matter and secondly develop their competence in a language other than the normal language of instruction." (ebd., S. 22).

Die Europäische Kommission (2003) legte fest, dass „every European citizen should have meaningful communication competence in at least two other languages in addition to his or her mother tongue." (S. 4). So werden Fremdsprachenkenntnisse vor dem Hintergrund der wirtschaftlichen und politischen Entwicklung Europas immer bedeutsamer (Burmeister 2002, Werner 2007, Wode u.a. 1999) und sollen für das Individuum zu verbesserter Berufsvorbereitung in einem globalisierten und vereinten Europa führen (Abuja 1999). Durch die Integration von Sprache und Inhalt soll der Nutzen und die Bedeutung für das Verwenden der Fremdsprache deutlicher und damit einhergehend eine größere Motivation zum Lernen der Fremdsprache erreicht werden (Wilhelmer 2008). Neben der Verbesserung von Fremdsprachenkenntnissen und kommunikativen Kompetenzen, wird der bilinguale Unterricht auch als effektiv angesehen: „If two things can be learned in the slot otherwise taken up by only one, this clearly saves time" (Dalton-Puffer / Smit 2007, zit. in Wilhelmer 2008, S. 45; vgl. Vollmer 2007).

Allerdings besteht oft eine „Diskrepanz zwischen fremdsprachlichen und kognitiven

Möglichkeiten der Lernenden in den Sachfächern" (Thürmann 2002), welche dazu führt, dass die Schülerinnen und Schüler weniger sagen als sie eigentlich kognitiv könnten (ebd.). Deshalb fordert Thürmann die Entwicklung einer für den bilingualen Sachfachunterricht spezifischen Unterrichtsmethodik. Dagegen stellt Rautenhaus (2000) fest, dass die „Schulpraxis der Theorie um Meilen vorausgeeilt" ist (zit. in Otten / Wildhage 2009). Als zentrale Frage einer möglichen Didaktik muss gefragt werden: „Is the language classroom integrating content or is the content classroom integrating language?" (Wilhelmer 2008, S. 5). Hellekjaer (1999) fordert eine „proper balance between language and subject-matter". Er ist weiterhin der Auffassung, dass der Sprachunterricht auf einem notwendigen Minimum gehalten werden sollte (ebd.). Dennoch muss man sich als Lehrkraft drei Anforderungen sprachlicher Art bewusst machen, die zu bewältigen und methodisch zu berücksichtigen sind:

(1) Verständnissicherung

(2) Entfaltung von Lernstrategien

(3) Einbeziehen von Sprachvergleichen bzw. Textentlastung (Neumann 2009).

Otten (1999) spricht in dem Zusammenhang von einem „funktionalen Einsatz von Fremdsprache(n) und Muttersprache(n)" und versteht darunter eine möglichst einsprachige Unterrichtsführung, bei welcher der Einsatz der Muttersprache zur Klärung von Sachverhalten und aus Gründen der Zeitersparnis erfolgt. Auf alle Fälle sollte die Fremdsprache nicht zum Gegenstand des Unterrichts werden, sondern lediglich als Vehikular- und Arbeitssprache dienen (Krechel 1999). Stattdessen sind Stützmaßnahmen und methodische Hilfen notwendig (ebd.). Bei der Auswahl des Stoffes sollte eine klare Schwerpunktsetzung vorgenommen und der Stoff entsprechend reduziert werden (ebd., vgl. Schlemminger 2008). Des Weiteren ist der Einsatz authentischer Materialien wie Realien, audiovisueller Medien, Fotos, Hörtexte, Zeitungsartikel etc. wichtig, um die Schülerinnen und Schüler zu motivieren (Krechel 1999). Das Verständnis der Fremdsprache sollte durch Illustrationen, das Umschreiben oder Paraphrasieren und notfalls auch durch Übersetzen unterstützt werden (Ministerium für Kultus, Jugend und Sport Baden-Württemberg 2006). Butzkamm (2000, zit. in Lose 2007) schlägt eine „Pendelstrategie" vor, bei der es im Ermessen der Lehrkraft liegt, welche Anforderungen dominieren sollen und zwischen inhaltlichem Kommunizieren und sprachlichem Üben hin und her „gependelt" wird.

Der bilinguale Unterricht bietet alle Erfolgsfaktoren zum Lernen einer Fremdsprache in der Schule: Er ermöglicht einen längeren und intensiveren Kontakt mit der Fremdsprache und bietet auf Grund des Sachfachkontextes auch thematische Vielfalt (vgl. Wode 2009).

Darüber hinaus erwerben die Schülerinnen und Schüler neben Basic Interpersonal Communicative Skills (BICS) auch Cognitive Academic Language Proficiency (CALP), ein abstraktes und formales Sprachlevel, welches kognitiv stärker herausfordernd, aber für akademische Studien unabdingbar ist (Bentley 2010). Dieses von Cummins entwickelte Konzept wurde später auch als content-obligatory language (im Gegensatz zu content-compatible language) bezeichnet (ebd.).

4.1. Mathematik als bilinguales Sachfach

Während man bei der Recherche zum bilingualen Unterricht für die Fächer Biologie, Geografie und Geschichte zahlreiche Materialien und Literatur finden kann, erstaunt die Abwesenheit desselben bei der Suche zum Mathematikunterricht umso mehr.

Begründet wird dies mit der Feststellung, Mathematik sei ungeeignet für den Einsatz im bilingualen Unterricht. Das Fach sei zu schwierig, um in der Fremdsprache unterrichtet zu werden (Küppers / Schmidt 2006). Am Ende könnten die Schülerinnen und Schüler weder rechnen noch besser Englisch sprechen (ebd.). Überhaupt werde die Sprache im Mathematikunterricht nicht benötigt, da ein Fokus auf Zahlen, Formeln und Rechenoperationen läge (Rolka 2004). Interkulturelles Lernen könne in diesem Fach nicht stattfinden (Küppers / Schmidt 2006), da es sich um eine neutrale Naturwissenschaft handele (Rolka 2004).

Dabei ist nach Hallet (2005, S. 6) ein Inhaltslernen von Sprache nicht möglich, Sprache ohne Inhalt ist demnach nicht denkbar. Kang und Pham (1995) sehen Mathematik als eigenständige Sprache mit spezifischem Vokabular und spezifischer Syntax an: "Mathematics has a unique register that student must ultimately learn." (S. 1). Auch Orando und Collier (1998, zit. in Dafouz / Llinares n.d., S. 183f.) stellen fest, dass „in reality, teachers and students employ a great deal of language in the math [...] teaching and learning process.", so dass man also durchaus davon ausgehen kann, dass auch im Mathematikunterricht Sprachenlernen stattfindet. Dass interkulturelles Lernen stattfinden kann, lässt sich schon an dem kleinen Beispiel der unterschiedlichen Datumsangabe in unserem Projekt nachweisen. Auch Whiteford (2009) ist überzeugt, dass „Mathematics can differ between cultures as much as language itself." (S. 280).

Tatsächlich ist Mathematik sehr geeignet als bilinguales Sachfach, da ein hohes Level an Sprachkompetenz nicht notwendig ist (Rolka 2004, S. 109) und es deshalb umso geeigneter als Einstieg in den bilingualen Unterricht angesehen werden kann (ebd.). Stattdessen ersetzen Symbole, Grafiken und Tabellen oft kognitiv anspruchsvolle Sprache (Wilhelmer 2008, S. 61). Viele der mathematischen Begriffe sind von lateinischem Ursprung und müssen sowieso gelernt werden (ebd.). Mathematik ist ein beständiges und stabiles Fach und durch seine aufeinander aufbauenden Kenntnisse für Wiederholungen geeignet (Schubnel 2009, S. 59).

Bilingualer Mathematikunterricht birgt auch hohes Motivationspotenzial, vor allem bei Schülerinnen und Schülern, die eher als „mathematisch unbegabt" angesehen werden müssen (Küppers / Schmidt 2006, S. 130): Der „common belief that people are either mathematical-scientifically or linguistically interested or talented" (Rolka 2004, S. 105) führe nämlich dazu, dass diese durch die Verwendung der Fremdsprache motiviert würden (Küppers / Schmidt 2006, S. 130).

Breidbach (2000) betont außerdem den Nutzen einer bilingualen Sprachkompetenz in diesem Fach: „Der Zugang zu einkommensträchtigen Tätigkeiten, die als zukunftsorientiert gelten und mit einem entsprechend sozialen Prestige belegt sind, scheint doch eher über ökonomisches, technisches und naturwissenschaftliches Wissen eröffnet zu werden." (S. 178).

Wie in allen anderen bilingualen Sachfächern wird auch im bilingualen Mathematikunterricht ein Fokus auf das Sachfach und nicht auf die Sprache gelegt. Dennoch muss der „Mathematiklehrer [...] sich stets bewusst sein, dass er drei Sprachebenen mit ihren jeweiligen Verständnisschwierigkeiten abtasten muss: L1 = Alltagssprache, L2 = Fremdsprache, L3 = Mathematische Fachsprache. [...] [Er] muss nicht nur den „Fehler" behandeln, sondern zunächst die Sprachebene identifizieren, aus welcher der Fehler (mutmaßlich) kommt." (Lorbeer n.d., S. 2). Die Lehrkraft muss bei zu anspruchsvollem mathematischen Niveau eventuell die Anforderungen in der Fremdsprache zurücknehmen (Schubnel 2009, S. 105) und sicherstellen, dass der Input sowohl von der sprachlichen als auch von der mathematischen Perspektive her verständlich ist (Wilhelmer 2008, S. 70f.). Auch im bilingualen Mathematikunterricht ist die Unterstützung des Verständnisses durch Hilfsmethoden unabdingbar: „Providing contextual support is the primary way to help make the language and content of lessons and texts comprehensible." (Kang / Pham 1995, S. 8)

5. Verlaufsplan

1st Lesson

Time	Phase / Function / Stage	Learning activities and Language	Social form	Media
0:00-0:10	Introduction	T: Hello, I am Miss Liebold. Today we want to do Maths in English. Can you all come to the front and sit down in a circle. (T gestures, students sit around circle)	circle	blue circle
		Let's start by introducing each other. My name is Miss Liebold. What is your name? (throw ball from student to student, hand out name tag)		ball, name tags
0:11-0:15	Silent Stimulus	Calendars in the middle T: What can you see? What is this? S: Kalender. T: Yes, it's a calendar (students repeat: Calendar) What is in a calendar? S: Wochen, Monate, Tage		different calendars
0:16-0:30	Development of topic 1	T: Let's start building a calendar. – What months do you know? S name months – T puts month in the middle T: Can you bring them into the right order?, T puts month names on blue circle		month cards (English)
		S: Ich hab im Juni Geburtstag… / T: When is your birthday? T: Oh, that is great. Your birthday is in June. On what day? (S names day – T repeats the ordinal) T gives student a present sticker to put on the month T: Now when is your birthday? (S repeats whole birthday)		present stickers
0:31-		T: What month does the year start with?		

14

Time	Stage	Content	Social form	Materials
0:45		S: January. T: January is the first month. What about February? (up to December, T puts month number) (let students repeat) T: So, what month was March? October?....		number cards
0:46-0:55	Consolidation 1	T: I brought you a memory – let's play it together. T shows how to play game		memory game (big)
		T: Now go in pairs and play together. Try saying the English words! S play memory game in pairs	pair work	memory game (small)
0:56-1:05	Development of topic 2	T: Please come back in the circle.	circle	
		T: Today is the 22nd of September – what day is tomorrow? Yesterday? What day is in a week? Two weeks' time? (T writes on board)		September calendar, blackboard
1:06-1:20	Consolidation 2	T: Now we want to find out for the other months.... Find your partner and take your month. (S find their month partner.) S work on month worksheet T: Let's come back into the circle. Can you present your month? S present month to other students L: How many days are there in January? (S put day stickers on blue circle)		month worksheet
				day stickers
1:21-1:30	Summary / End of lesson	T: Let's learn a song together!		The Month song

2nd Lesson

Time	Phase / Function / Stage	Learning activities and Language	Social form	Media
0:00-0:15	Activation of previous know-ledge	T: Let's repeat the song from last lesson... (sing load, quiet, faster, slow) I brought the names of the month on cards. What month is this? (name, number) (T hands out month cards) T: Who has...? (S holds month card up) T: Good! Let's sing the song again now! Show me your card when we say your month! (S show card at the right time during song)	class	The Month song, month cards
0:16-0:25		T: Let's line up. – Stellt euch nach eurem Geburtstag auf – ohne dabei zu reden! T: Let's check. (S name their birthdays.)		
		T: Okay, now we'll go into groups – 4 students (S get flag sticker)		flag stickers
0:26-0:30	Development of topic 3	T: Sit down at a table together... Let me introduce Rudi to you :) Rudi is a German student. He is very lazy and doesn't like to go to school... So he thought he could go to a different country. Can you help Rudi find a country where he has lots of holidays? (T gives out holiday worksheets in envelope)		Lazy Rudi, holiday worksheets (envelope with calendar, holidays Germany, holidays foreign country, task)
0:31-1:00	Consolidation 3	Group work		
1:01-1:20	Presentation	S present their group, put number of holidays on world map T: Thank you. That was great! Now where do you think Rudi would go to?! (compare holidays)		world map
1:21-	Feedback	Feedback		Smiley/Frowny

1:30		- Wie war es für euch Mathe auf Englisch zu machen? - Wie hat euch das Thema gefallen? - Hat euch die Gruppenarbeit Spaß gemacht? - Hat euch die Erarbeitung im Kreis gefallen? Möchte noch jemand etwas sagen? T: Es war toll, mit euch zu arbeiten. Vielen Dank fürs Mitmachen! (The Month song?)

6. Portfolio zum Auslandssemester

From the beginning of my studies in Germany, it was clear to me that I would go abroad and study there as part of it. I did not just make this decision because the curriculum of my field of studies requires attending university in an English speaking country for one semester, but also because I had just returned from a two-year stay in the USA where I had worked as an au pair. I missed the USA and I could not wait to be abroad for a period of time again.

As soon as it was clear that I would study at Central Connecticut State University, I started preparing my stay abroad carefully. I bought flight tickets for my trip at the end of August 2010 as well as for my return in June 2011. I also scheduled an appointment at the American Embassy in Frankfurt/Main where I wanted to apply for a visa for my stay. In order to do that, I obtained form DS-2019 in the mail. The university had filled it out and signed it. I turned it in together with the application forms and additional documents like passport photos and proof of paid SEVIS fee. Apart from that, I was also required to provide documentation that I had funds to support my stay abroad and that I would return to my home country after the visa expired. I received my passport with the attached J1 visa about two weeks after my visit to the embassy at the end of July.

I also thought about my future life in the USA. I decided to look for a room off-campus because the rent for dorm rooms was pretty high (about $2000 per semester) and could only be rented in combination with a meal plan which cost about the same amount of money. Not only did this seem really expensive to me, I also worried that I would have to share my room with someone I did not get along with very well. Because I was not sure about the quality of food on campus either, I chose to look for a private room close to the university. I looked at classified ads from the New Britain area and found many interesting offerings on the website Craigslist as well as Facebook Marketplace. Eventually, I found a functional room in a house which was inhabited by several young people only two blocks away from campus. The landlord did not mind that I would not pay security and rent until after my arrival and even suggested that I could sleep on the couch for free during the last week of August while my room was still rented to someone else. This way, I had a place to sleep when I arrived in the USA and did not have to stay in a hotel in the meantime.

Another thing I considered was purchasing a car for the time that I lived in Connecticut. I knew from experience that life in the US without a car can be hard and tricky. I also spoke with the exchange student who studied in New Britain the year before and she suggested buying a car as well. I liked the idea of being independent and decided to look for a pre-owned car right after my arrival in the US.

I selected my classes while I was still in Germany. I wanted to take classes that would count towards my credits in Germany but also classes that interested me personally. Therefore, I registered for an Introduction to Philosophy class which turned out to be difficult because I noticed in class and especially during exams that even though I was fluent in English, I still reached a point that was linguistically demanding. I also took an Introduction to American Studies class although I had taken one in Germany already. I was interested in the American perspective on the topic. I attended a French Introductory class as well because I always wanted to learn this language but had not yet had the opportunity in Germany. My favorite class was the TESOL Methods class. We talked about different ways of teaching English as a second or foreign language. I enjoyed this class very much because we worked hard and were productive. Almost all of my classmates had learned English as their second or a foreign language and therefore it was easy to imagine the students' point of view. As part of our requirements for this class, we had to sign up to be a conversation partner and report about these meetings for several weeks. In addition, we observed two classes in the Intensive English Language Program (IELP) at Central Connecticut State University. The semester finished with the teaching of our own individual language lesson to the students of the IELP.

Because of this class, I got to know the IELP. It is a program designed to teach students from all over the world English in an intensive all-day manner. The students come to university full-time and focus on reading/writing and listening/speaking in separate classes. They are assigned to different levels of difficulty according to their language level. It is possible to switch between different levels at the end of each eight-week session.

My conversation partner from Egypt knew only very basic English when we met the first time. I found out more about the program and met other students who studied English as well. Before I came to my American university, I was promised an assistant teacher position in the German department of the university. After I arrived, however, I found out that the department was undergoing major personnel cutbacks, and so I could not become a tutor there. Because I had more free time than first expected, I was able to extend the meetings with my conversation partner over the number of sessions required for my TESOL Methods class. In fact, I helped my conversation partner to improve his English until I returned back to Germany. For the most part, I had contact with students from Arabic speaking countries and experienced how students struggled with word order, pronunciation and vocabulary words. They also had to learn new letters and get used to reading from left to right. It was not until I found out more about their mother tongue that I understood some of the mistakes the students made. I discovered that in Arabic, at least in some dialects, "be" is omitted, making me realize why students did not use any form of "be" in their English sentences either. Similar problems happened when looking at the word order in sentences. I met another tutor who studies Teaching English as a Second Language and whose mother tongue is English because he is American. We had a lot of interesting conversations about the quality of the IELP classes.

These conversations as well as encounters with other students made me realize that an English language program has to do more than just teach the language. First of all, the teacher must put himself/herself in the student's place. To do that, it is necessary to find out more about the student's native language and at least understand basic structures of the language to avoid misinterpreting the student's mistakes as stupidity or ignorance but instead to realize the errors as transfer from the native language and fight them. It is also helpful if the teacher has experiences in acquiring a foreign language. This can help to understand the learning process that takes place. It will also help to better understand errors and misunderstandings that always go along with it. Unfortunately, I found out that some teachers in the IELP did not share this attitude and therefore the quality of some of the lessons was not satisfying. I had the feeling that there were no mandatory standards. For example, it would have been preferable to have a list of vocabulary words for each language level that should be acquired. During homework, I noticed more than once that students had to memorize pages and pages copied out of picture dictionaries. These often included very detailed and unnecessary vocabulary words. It seems best to me to challenge the selection of vocabulary words and ask which words are absolutely necessary for the students to acquire. This will be easier to accomplish if you, as a teacher, have been in a foreign country for some period of time and had to master everyday situations.

Although I could not teach as originally intended, this stay abroad helped my professional development and I learned a lot during these months. In addition to attending classes and being a conversation partner, I also worked as a cashier in the university bookstore. I loved this job because it gave me the opportunity to further improve my English language skills, meet new people and have interesting conversations. I think these experiences will be helpful for future employment in Germany as well.

Of course, I also took the opportunity to travel a lot. I went on a few weekend trips to Washington D.C., Boston, and New York City. I spent Thanksgiving at my former host family's home in Michigan, Christmas with friends near Hartford, and celebrated New Year's in Michigan as well. I enjoyed spring break and warm temperatures in Florida while it snowed in Connecticut and met an old friend after several years in Toronto, Canada. I met many people who I call friends now and hope to stay in contact with for a long time.

During my stay abroad I encountered several situations that brought me to the brink of despair and that I had to manage all by myself. I learned during that time that being strong-willed can bring you far. It is scary sometimes when you are far away from home and all by yourself, and you have to deal with problems that seem insurmountable. It is a nice feeling when "everything works out" in the end and I started to adopt this American attitude that makes you see things much more realistically and much less terrifying.

Studying in the US is much more intensive than it is in Germany. During the semester there is much more homework to do and classes have to be prepared with more depth. It is expected that students have worked with the material. The course grade usually consists of a few small papers and essays. In contrast to that, in Germany students only have to write a term paper at the end of the semester, during the break. I noticed that I got much more involved in my classes in the US because I had to work hard outside of class as well. I never had a similar experience while studying in Germany. I took five different classes during the second semester – American Sign Language, Algebra, Medial English, Children's Literature and French II. As a result, the coursework was enormous, so I spent my Fridays off doing my load of homework.

Campus life is more diverse than in Germany. There are numerous clubs and organizations as well as sports teams. It seems like these clubs are supported financially because most of the events are free for students. On the CCSU campus there is the student center, a building where students can go between and after classes. There is a small cafeteria in the building, many places to sit back, relax, and learn on three levels, as well as a room with pool tables, game consoles, and board games. There are also many conference rooms in which different events take place. The bookstore is here too. Students cannot only buy textbooks but also CCSU apparel, some groceries and stationary. I spent a lot of my free time in the student center, going to different events. I especially liked Thursday nights when there was a free movie and a theme party that was always fun to go to. There was a Halloween

Party, a Hispanic Party, a party with a fortune teller and one with a ghost hunter. Looking back I think these activities made life as a student in the US special because other than that, life at an American university is similar to life as a student in Germany.

It turned out to be a good decision to live off campus. I met a few people who were not very happy with their dorm room and / or with food on campus and who moved off campus during the course of the semester. Because I had a room in a shared house, I met people who did not go to university anymore. It was a good experience. I also did not regret buying a car – I drove 20,000 miles in ten months (I walked to campus).

Overall, my time at an American university was successful. Especially because of the personal connections I made, my year in New Britain was an unforgettable and priceless experience.

Literaturverzeichnis

Abuja, Gunther (1999). *Die Verwendung einer Fremdsprache als Arbeitssprache: Charakteristika 'bilingualen Lernens' in Österreich.* In: http://zif.spz.tu-darmstadt.de/jg-04-2/beitrag/abuja2.htm (18.08.2011)

Badertscher, Hans / Bieri, Thomas (2009). *Wissenserwerb im Content and Language Integrated Learning.* Basel.

Bentley, Kay (2010). *The TKT Course – CLIL Module.* Cambridge.

Breidbach, Stephan (2000). *Bilinguale Didaktik zwischen allen Stühlen? Zum Verhältnis von Fremdsprachendidaktik und Sachfachdidaktiken.* In: Bach, Gerhard / Niemeier, Susanne (Hrsg.). Bilingualer Unterricht. Grundlagen, Methoden, Praxis, Perspektiven. Zweite überarbeitete und erweiterte Auflage. Frankfurt am Main, S. 173-184.

Burmeister, Petra (2002). *Bilingualer Unterricht – Ergebnisse aus einem Forschungsprojekt.* In: Hilligus, Annegret Helen / Rinkens, Hans-Dieter / Friedrich, Claudia (Hrsg.). Europa in Schule und Lehrerausbildung. Entwicklungen – Beispiele – Perspektiven. Münster, S. 51-58.

Dafouz Milne, Emma / Llinares García, Ana (n.d.). *Das Lehren einer Fremdsprache über Inhalte.* In: http://www.europa-bilingual.net/part1_d/Dafouz-Llinares-d.pdf (16.08.2011)

Commission of the European Communities (2003). *Communication from the Commission to the Council, the European Parliament, the Economic and Social Committee and the Committee of the Regions. Promoting Language Learning and Linguistic Diversity: An Action Plan 2004 – 2006.* In: http://ec.europa.eu/education/doc/official/keydoc/actlang/act_lang_en.pdf (31.08.2011)

Eurydice (Hrsg.) (2006). *Content and Language Integrated Learning (CLIL) at School in Europe.* In: http://www.mp.gov.rs/resursi/dokumenti/dok36-eng-CLIL.pdf (02.09.2011)

Görke, Winfried (2011). *Datum und Kalender. Von der Antike bis zur Gegenwart.* Heidelberg.

Hallet, Wolfgang (2005). *Sprachliches Lernen im Bilingualen Unterricht.* In: Der Fremdsprachliche Unterricht Englisch CLIL November 2005, S. 2-8.

Hellekjaer, Glenn Ole (1999). *Easy does it: Introducing Pupils to Bilingual Instruction.* In: http://zif.spz.tu-darmstadt.de/jg-04-2/beitrag/hellek1.htm (18.08.2011)

Kang, Hee-Won / Pham, Kien T. (1995). *From 1 to Z: Integrating Math and Language Learning.* In: http://www.eric.ed.gov/PDFS/ED381031.pdf (14.09.2011)

Kidsnet. *Der Kalender.* In: http://www.kidsnet.at/sachunterricht/kalender.htm (23.09.2011)

Krechel, Hans-Ludwig (1999). *Sprach- und Textarbeit im Rahmen von flexiblen bilingualen Modulen.* In: http://zif.spz.tu-darmstadt.de/jg-04-2/beitrag/krechel1.htm (18.08.2011)

Küppers, Almut / Schmidt, Dietlinde (2006). *Mit der Mathematik rechnen! Zahlenzauber im bilingualen Unterricht?* In: Küppers, Almut / Quetz, Jürgen (Hrsg.). Motivation Revisited. Berlin, S. 125-135.

Lorbeer, Werner (n.d.). *Gibt es eine Didaktik bilingualen Mathematikunterrichts?* In: http://www.mathematik.uni-muenchen.de/~didaktik/doc/kolloquium/GibtEsDidaktik.pdf (16.08.2011)

Lose, Jana L. (2007). The language of scientific discourse: Ergebnisse einer empirisch deskriptiven Interaktionsanalyse zur Verwendung fachbezogener Diskursfunktionen im bilingualen Biologieunterricht. In: Caspari, Daniela / Hallet, Wolfgang / Wegner, Anke / Zydatiß, Wolfgang (Hrsg.). *Bilingualer Unterricht macht Schule.* Frankfurt am Main, S. 98-107.

Ministerium für Kultus, Jugend und Sport Baden-Württemberg (Hrsg.) (2004). *Bildungsplan Grundschule.* In: http://www.bildung-staerkt-menschen.de/service/downloads/ Bildungsplaene/Grundschule/Grundschule_Bildungsplan_Gesamt.pdf (19.08.2011)

Ministerium für Kultus, Jugend und Sport Baden-Württemberg (Hrsg.) (2006). *Realschule – Bilingualer Unterricht.* In: http://www.schule-bw.de/schularten/realschule/bilingual/ publikationen/handreichung.pdf (16.08.2011)

Neumann, Ursula (2009). *Der Beitrag bilingualer Schulmodelle zur Curriculuminnovation.* In: Gogolin, Ingrid / Neumann, Ursula (Hrsg.). Streitfall Zweisprachigkeit – The Bilingualism Controversy. Wiesbaden, S. 371-331.

Otten, Edgar (1999). *Nachdenken über den funktionalen Einsatz von Fremdsprache(n) und Muttersprache(n) in der inhaltsbezogenen Arbeit.* In: http://zif.spz.tu-darmstadt.de/jg-04-2/beitrag/otten5.htm (18.08.2011)

Otten, Edgar / Wildhage, Manfred (2009). *Content and Language Integrated Learning. Eckpunkte einer "kleinen" Didaktik des bilingualen Sachunterrichts.* In: Wildhage, Manfred / Otten, Edgar (Hrsg.). Praxis des bilingualen Unterrichts. 3. Auflage. Berlin, S. 12-45.

Rolka, Katrin (2004). *Bilingual Lessons and Mathematical World Views – A German Perspective.* In: http://www.emis.de/proceedings/PME28/RR/RR189_Rolka.pdf (13.09.2011)

Sawelski, S. F. (1977). *Die Zeit und ihre Messung.* Leipzig.

Schlemminger, Gérald (2008). *Prolegomena eines oberrheinischen Modells zum bilingualen Lehren und Lernen.* In: Schlemminger, Gérald (Hrsg.). Erforschung des Bilingualen Lehrens und Lernens. Forschungsarbeiten und Erprobungen von Unterrichtskonzepten und – materialien in der Grundschule. Erlangen, S. 13-57.

Schubnel, Yves (2009). *Bilingualer Mathematikunterricht. Ein Beitrag zu einem zusammenwachsenden Europa.* In: http://opus.bsz-bw.de/phfr/volltexte/2010/381/ pdf/Dissertation.pdf (15.08.2011)

Sekretariat der Ständigen Konferenz der Kultusminister der Länder in der Bundesrepublik Deutschland (2006). *Bericht „Konzepte für den bilingualen Unterricht" – Erfahrungsbericht und Vorschläge zur Weiterentwicklung"*. In: http://www.kmk.org/fileadmin/veroeffentlichungen_beschluesse/2006/2006_04_10-Konzepte-bilingualer-Unterricht.pdf (28.08.2011)

Thürmann, Eike (2002). Eine eigenständige Methodik für den bilingualen Sachfachunterricht? In: Bach, G. / Niemeier, S. (Hrsg.). Bilingualer Unterricht. Frankfurt am Main, S. 75-93.

Vollmer, Helmut J. (2007). Bilingualer Sachfachunterricht als Inhalts- und als Sprachlernen. In: Caspari, Daniela / Hallet, Wolfgang / Wegner, Anke / Zydatiß, Wolfgang (Hrsg.). *Bilingualer Unterricht macht Schule*. Frankfurt am Main, S. 51-73.

Werner, Bettina (2007). *Entwicklungen und aktuelle Zahlen bilingualen Unterrichts in Deutschland und Berlin*. In: Caspari, Daniela / Hallet, Wolfgang / Wegner, Anke / Zydatiß, Wolfgang (Hrsg.). Bilingualer Unterricht macht Schule. Frankfurt am Main, S. 19-28.

Whiteford, Tim (2009). *Is Mathematics a Universal Language?* Teaching Children Mathematics, December 2009/January 2010, S. 276-283.

Wilhelmer, Nadja (2008). *Content and Language Integrated Learning (CLIL). Teaching Mathematics in English*. Saarbrücken.

Wode, Henning (2009). *Frühes Fremdsprachenlernen in bilingualen Kindergärten und Grundschulen*. Braunschweig.

Wode, Henning / Burmeister, Petra / Rohde, Andreas (1999). *Verbundmöglichkeiten von Kindergarten, Grundschule und Sekundarstufe I im Hinblick auf den Einsatz von bilingualem Unterricht*. In: http://zif.spz.tu-darmstadt.de/jg-04-2/beitrag/wode2.htm (17.08.2011)